DL019849
£20.00
28/11/03

WOLVERHAMPTON COLLEGE

D0264054

Health and safety in construction

Guidance for construction professionals

John Barber

T¹ ThomasTelford

Published for the Institution of Civil Engineers by Thomas Telford Publishing, Thomas Telford Ltd, 1 Heron Quay, London E14 4JD.
URL: http://www.thomastelford.com

Distributors for Thomas Telford books are
USA: ASCE Press, 1801 Alexander Bell Drive, Reston, VA 20191-4400
Japan: Maruzen Co. Ltd, Book Department, 3–10 Nihonbashi 2-chome, Chuo-ku, Tokyo 103
Australia: DA Books and Journals, 648 Whitehorse Road, Mitcham 3132, Victoria

First published 2002
A catalogue record for this book is available from the British Library

ISBN: 0 7277 3118 1
© Thomas Telford Limited, 2002

This Guidance has been written, so far as possible, to avoid gender-specific pronouns. The pronouns 'he', 'his', 'him' or 'himself' have been used, where necessary, in order to quote accurately from legislation or conditions of contract, e.g. referring to the 'designer' or the 'planning supervisor'. The masculine pronouns are used in this context, as a matter of legal interpretation, to refer generally to a legal 'person' which can be either an individual, male or female, or an organisation that is a legal entity.

All rights, including translation, reserved. Except as permitted by the Copyright, Designs and Patents Act 1988, no part of this publication may be reproduced, stored in a retrieval system or transmitted in any form or by any means, electronic, mechanical, photocopying or otherwise, without the prior written permission of the Publishing Director, Thomas Telford Publishing, Thomas Telford Ltd, 1 Heron Quay, London E14 4JD.

This book is published on the understanding that the author is solely responsible for the statements made and opinions expressed in it and that its publication does not necessarily imply that such statements and/or opinions are or reflect the views or opinions of the publishers. While every effort has been made to ensure that the statements made and the opinions expressed in this publication provide a safe and accurate guide, no liability or responsibility can be accepted in this respect by the author or publisher (see also paragraph 1.9).

Typeset by Alex Lazarou, Surbiton, Surrey
Printed and bound in Great Britain by Bell & Bain Ltd, Glasgow

Foreword

WOLVERHAMPTON COLLEGE
LEARNING CENTRE

690.22 BAR WR

8785172

2 DEC 2009

People are killed in construction-related activities, on average, every week of the year. Sometimes, in bad years, three are killed each week, but in 'better' years there is still, at least, one fatality on average per week.

What can be done? Possibly there is no totally effective solution. Where construction work is carried out, there will be risk. But many of the problems occur on small projects, not on major works. The real problem is often a lack of awareness of the hazards and risks involved, coupled with insufficient determination to control the risks. The challenge is not how to make some projects super safe, but how to ensure acceptable standards on all projects.

Some of the problems can be solved on site but we must also do more to reduce risk at source. Designers create many of the risks, contractors can mostly only manage the risks, while the workers have to endure them.

Clients must learn more about their responsibilities and think about the risks. They have previous knowledge of the site — they can and should provide information. They can also adjust their requirements to allow, so far as possible, the works to be designed and specified to be constructed, repaired and demolished safely and without unacceptable risk to health.

Designers 'design in' risk, although not deliberately. Designers must think how to 'design out' risk, for example, by eliminating deep trenches, fragile materials and the need to work at heights.

Construction professionals are involved at all these stages. We need to work together to ensure consistently high standards. It is hoped that this publication, giving clear guidance on the responsibilities of construction professionals for health and safety in construction, will help to promote a measured and effective response to the challenge.

John Fisher MBE
Chairman, Health and Safety Board 1999–2001

WOLVERHAMPTON COLLEGE

Acknowledgements

This document has been produced under the joint aegis of the Health and Safety Board (Chairman, John Fisher MBE) and the Advisory Panel on Legal Affairs (Chairman, Peter Chapman) of the Institution of Civil Engineers. It replaces the separate guidance documents, *The Management of Health & Safety in Civil Engineering* (1995) and *ICE Legal Notes — Health and Safety in Construction* (1991). The text and case studies have been written by John Barber, who is a consulting engineer and lecturer at the Centre of Construction Law, King's College London. Detailed comments on drafts have been received from (in order of receipt):

    Sir Alan Muir Wood
    Professor Donald Bishop CBE
    Donald Lamont
    Dr John Anderson
    Gillian Birkby
    David Lambert
    Roger Ball
    Jeremy Winter

Co-ordination and support have been provided by:

    Drick Vernon of the Institution of Civil Engineers

The extracts for Rules for Professional Conduct in Appendix A are reproduced by kind permission of the Institution of Structural Engineers, the Royal Institute of British Architects, the Royal Institution of Chartered Surveyors and the Chartered Institute of Building. Appendix B is reproduced by kind permission of the Royal Academy of Engineering. Appendix C is reproduced from a paper by Edmund Terry and Simon Dean, published by the European Construction Institute.

Note on the case studies

The case studies used to illustrate the text are mostly cases where the facts are in the public domain following prosecutions. The others relate to experience, but the facts have been varied both to conceal identities and to make points more clearly. The prevalence in the case studies of mistakes rather than success is not a reflection of what happens generally in construction. The point is that, in engineering for safety, we can learn more from mistakes than from success.

Please note: Updates to the text will be posted on the ICE website at www.ice.org.uk.

Glossary

ACoP	Approved Code of Practice
CDM	Construction (Design and Management) Regulations 1994 (as amended 2000)
CHSWR	Construction (Health, Safety and Welfare) Regulations 1996
CIOB	Chartered Institute of Building
COMAH	Control of Major Accident Hazards Regulations 1999
COSHH	Control of Substances Hazardous to Health Regulations 1999
EU	European Union
HSC	Health and Safety Commission
HSE	Health and Safety Executive
HSWA	Health and Safety at Work etc. Act 1974
ICE	Institution of Civil Engineers
IStructE	Institution of Structural Engineers

LLP	Limited Liability Partnership
LOLER	Lifting Operations and Lifting Equipment Regulations 1999
MEWP	Mobile elevating work platform
MHSWR	Management of Health and Safety at Work Regulations 1992 and/or 1999
NEC	*NEC Engineering and Construction Contract*
PPE	Personal Protective Equipment Regulations 1999
PUWER	Provision and Use of Work Equipment Regulations 1998
RIBA	Royal Institute of British Architects
RICS	Royal Institution of Chartered Surveyors
RIDDOR	Reporting of Injuries, Diseases and Dangerous Occurrences Regulations 1995
SCOSS	Standing Committee on Structural Safety
UK	United Kingdom

Contents

1 Introduction

1.1   All civil engineers and other construction professionals concerned with the construction process have both professional and legal duties to take care, not only of their own health and safety at work, but the health and safety of others who might be put at risk by their acts or omissions.

1.2   The professional duties of members of the construction professional bodies are stated in their respective Rules for Professional Conduct. This publication is intended primarily as guidance from the Institution of Civil Engineers (ICE) to its members on the responsibilities of the professional civil engineer for health and safety, having regard to the *ICE Rules for Professional Conduct*. However, the guidance should be equally relevant to other construction professionals, and therefore it is addressed to construction professionals generally.

1.3   Some legal duties apply directly to the individual, but most apply primarily to the individual's employer or to the management of a project, and only indirectly to the individual. The duties arise through:

- civil liability for personal injury or damage to property or the environment

- the framework of duties laid down by statute, as interpreted by the courts, enforceable by criminal penalties and/or enforcement notices
- contractual obligations and powers, both express and implied.

1.4 Professional and legal duties should be compatible. As a general principle, the professional duty incorporates the legal duty. Construction professionals therefore need to be aware of the existence and scope of the legal duties, and the standards of care required, to ensure that they, and their employers, discharge the duties and thereby avoid civil and/or criminal liability for breach of the duty. Construction professionals should ensure that they carry out their own work to discharge these duties, that persons under their control also do so and, if appropriate, that their employers and/or clients are advised of relevant duties. (For consistency with health and safety legislation, the term 'employer' will be used throughout this Guidance to denote an organisation which employs individual employees, and not in the sense in which it is used in standard forms of construction contract, unless quoting directly.)

1.5 Construction professionals need to keep up to date with relevant legislation and regulations (what is new, amended or revoked) and interpretations of legislation and regulations by the courts, and to be aware of the impact of changes. For example, there were major shifts during the 1990s towards:

- regulations being framed as goal-setting rather than prescriptive, following the Cullen Report (1990)
- risk assessment being adopted as a mandatory requirement and a cornerstone of health and safety strategy following

the EU Framework Directive (1989) and the Management of Health and Safety at Work Regulations 1992

- the imposition of non-delegable duties on clients following both the decision of the House of Lords in *R* v *Associated Octel* (1996) and the CDM Regulations 1994.

1.6   In addition, as part of their professional duties, construction professionals have to take a broader and longer-term perspective. Health and safety in construction is concerned primarily with promoting the health and safety of those involved in the construction process or of the public who might be affected by the construction process, but construction professionals need also to ensure the long-term safety, durability and environmental performance of projects. We need to learn and share any lessons learnt from accidents that do occur: that is, how engineering and management progress. These additional concerns could be a source of conflict with health and safety law. Therefore, the aim of this Guidance is not only to explain in outline the current requirements of the law as regards health and safety, but also to suggest how potential conflicts between the professional and legal duties should be reconciled by construction professionals.

Case Study 1

Engineers designing piled foundations decided against driven piles and changed to bored cast-in-situ piles, to eliminate the health risk of noise. However, the bored cast-in-situ piles suffered necking in the soft clay strata and failed under load. This would not have occurred with the driven piles. The structure eventually had to be demolished and rebuilt.

1.7   Health and safety concerns range from welfare provisions for operatives, through good housekeeping and safe working practices on site, to technical matters, which may range from the commonplace to the limits of current knowledge. Construction professionals should not lose sight of the importance of all levels of safety concerns. A sense of balance should be maintained in regard to the thought, effort and resources to be applied to different aspects.

1.8   The achievement of safety depends on vigilance, observation and understanding, combined with a systematic approach. Formal systems and procedures can be valuable in complementing and assisting vigilance, observation and understanding, but they are not a substitute. They should not be allowed to defeat the main purpose, for example, by obscuring important matters with a proliferation of minor ones. The foundation of a systematic approach should be the development of safe habits of working and sound standard practices.

Case Study 2

The HSE Report (2000) into the collapse of the NATM (New Austrian Tunnelling Method) tunnels at Heathrow observed:

The pro forma approach [to risk assessment] kept the focus on routine worker safety. It did not encourage the strategic identification of high-level engineering and management issues essential to the success of NATM, such as major accident hazard events and their prevention.

1.9   This Guidance is based on the law in the UK. (The relevant criminal law is the same throughout Great Britain. Separate but similar legislation exists in Northern Ireland, the Channel Isles and the Isle of Man. There are minor differences, mainly of terminology and relevant statutes, between English and Scots law on civil liability.) **The Guidance is not intended as a comprehensive statement of the law, nor as a substitute for specific legal advice. Construction professionals should consult the full text of the statutory and regulatory sources, and obtain legal advice where appropriate.** Updates will be posted on the ICE website at www.ice.org.uk.

1.10   UK health and safety law generally only applies within the UK and to certain offshore work. When working in other jurisdictions, construction professionals should take steps to ascertain and comply with the local law. Other EU countries are likely to have similar health and safety legislation, since it will be derived from common EU Directives.

2   Professional duties

2.1   Extracts from Rules for Professional Conduct of other relevant professional institutions relevant to health and safety are set out in Appendix A. The *ICE Rules for Professional Conduct* (as amended 1999) state, so far as relevant:

> *Rule 1. A member shall discharge professional responsibilities with integrity and shall not undertake work in areas in which the member is not competent to practise.*
>
> *Rule 3.   A member shall have full regard for the public interest, particularly in relation to the environment and to matters of health and safety.*
>
> *Rule 11.   A member shall, consistent with safety and other aspects of the public interest, endeavour to deliver to the employer or client cost effective solutions. A member shall not comply with any instruction requiring dishonest action or the disregard of established norms of safety in design and construction.*
>
> *Rule 16.   If requested to comment on another engineer's work a member shall act with integrity...*

2.2   The Rules are directed at the respective institutions'
        members as individuals, since the institutions have no
        organisations as members. However, it is implicit that
        members should also discharge their professional duties
        through any organisation or team in which they play a role,
        particularly if they are directors, partners or managers.

2.3   When construction professionals are called upon to act as
        experts on safety related matters in court proceedings,
        whether civil or criminal, they should recognise their
        responsibility to give a balanced, objective and thoughtful
        opinion.

3 Civil liability

3.1   A person (which, in law, can mean not only an individual but also a limited company, a public authority or, effectively, a partnership) may be liable to pay compensation (known as 'damages') to a party who has suffered injury caused by a breach of a legal duty owed by the person to the injured party. Legal duties of care are owed towards specific people or classes of people, and the nature of the duty and consequent liability will depend on the relationship. Liability generally only arises, in the absence of a contractual relationship, when injury or damage has actually been suffered. (The Contracts (Rights of Third Parties) Act 1999 now permits liability in some circumstances, under English law, for breach of a contractual duty towards a person who was not a party to the contract. This possibility also exists under Scots law.)

3.2   In particular, legal duties of care are owed by employers towards their employees. These may give rise to liabilities to an employee injured as a result of a breach of:

- the employer's common law duty of care towards the employee, in particular a duty to provide a safe system of work
- a specific duty imposed by statute insofar as such duty gives rise to civil liability.

3.3   Duties of care may exist towards persons other than direct employees, giving rise to liability for injury, loss or damage suffered by such persons caused by a breach of the duty. Such duties include:

- a common law duty of care in carrying out work
- a common law duty of care in providing advice as a professional or specialist organisation
- the duty of care imposed on occupiers by the Occupier's Liability Acts 1957 and 1984 or, in Scotland, the Occupier's Liability (Scotland) Act 1960
- specific duties imposed by statute insofar as such duties give rise to civil liability.

3.4   Prior to the Health and Safety at Work etc. Act 1974 (HSWA), a breach of a duty imposed by health and safety statutes commonly gave rise to civil liability as well as criminal liability. The courts inferred a legislative intent to impose the duty for the benefit of the persons at risk. However, the enactment of the HSWA coincided with a change in legislative practice. Parliament moved to spelling out in each statute whether duties imposed by the statute were intended to confer entitlement to compensatory damages for injury caused by a breach. In accordance with this trend, Section 47 of the HSWA states that there is no civil liability for breach of the general duties, nor for breaches of the duties imposed by health and safety regulations except insofar as the regulations provide otherwise. The exceptions are few.

3.5   The general duties under HSWA are, in fact, deliberately similar to the duties of care giving rise to civil liability at common law. However, there are differences, particularly because the purposes of civil and criminal law are different. Civil law is concerned with compensating injury. Criminal law is concerned with punishing fault. Once it has been established that there has been a breach of a duty of care, civil law is only concerned with the injury suffered, its consequences and whether or not it has been caused by the breach. Civil law is not concerned with the degree of fault (except in cases of contribution where liability is to be apportioned between several defendants, or contributory negligence where it is to be apportioned between defendant and claimant). The criminal law on health and safety is, or should be, concerned with the degree of fault, rather than the severity or consequences of the injury caused (if any). (An exception to this principle would be the proposed offences of corporate killing and killing by gross negligence, where the offences envisaged by the draft Bill depend on causing death by gross negligence.)

3.6   Individuals may be pursued directly for damages if they personally owed and have been in breach of a duty of care. (The law in this area is subject to review and construction professionals should be alert for details of developments.) However, civil legal proceedings will usually only be brought against the employer because it is the employer who is insured. An employer (indemnified by its insurers) will normally be held 'vicariously' liable for the acts and omissions of its employees acting in the course of their employment. Individuals who practise as partners in a partnership may find themselves personally liable because they are sued personally in order to access the partnership's insurance. (New limited liability partnerships (LLPs), formed under the Limited Liability Partnerships Act 2000, should avoid this danger as regards work undertaken by the LLP.)

Case Study 3

A firm of consulting engineers practising as a partnership was sued in negligence for £9 million after temporary works, which they had designed, collapsed and a number of people were seriously injured. The firm had personal insurance cover of £5 million, subject to an excess of £200,000. Insurers contested liability in the name of the firm, but the judge awarded damages of £5.8 million. Insurers paid up to the limit of cover, but the partners, including two retired partners, were personally liable for £800,000, in addition to the excess of £200,000, although only one partner had been directly involved in the design.

3.7   One organisation can also find itself effectively liable for the work of another, if it was under a duty to supervise that other organisation. The share of blame in such a situation may only be a small percentage, but the doctrine of joint and several liability can mean that the supervisor will pick up the full liability if the supervisee has become insolvent or cannot be pursued, e.g. is outside the jurisdiction. (This only applies to civil liability, not to criminal liability.) Liability can also arise from failure to warn of a dangerous situation. Construction professionals may well have a duty to warn of defects or deficiencies that they observe in the work of others. For example, a contractor may have a duty to warn the employer or designer if he considers that the design that he is being asked to construct is dangerous or otherwise deficient. Similarly, a duty may arise for a construction professional to warn a contractor who is observed to be carrying out work in a dangerous or otherwise inappropriate manner. A construction professional who fails to warn of such defects or deficiencies may become liable to parties who suffer loss or injury as a result. The liability may arise in tort (probably only where there is physical injury or damage) or in contract, whether by an express or implied term.

3.8   An organisation or other legal entity may be liable to persons injured while on its property, by virtue of being an 'occupier' under the Occupier's Liability Acts. A specific duty of care is imposed on an occupier by the Acts. This liability is subject to the 'independent contractor defence'. The Occupier's Liability Act 1957 provides a special defence that the occupier is not to be held liable if the danger was due to the faulty execution of any work of construction, maintenance or

Case Study 4

In *AMF International Ltd* v *Magnet Bowling Ltd* (1968) the judge had to interpret the Occupier's Liability Act 1957. He stated:

In the case of the construction of a substantial building...I should have thought that the building owner, if he is to escape subsequent tortious liability for faulty construction, should not only take care to contract with a competent contractor...but also to cause that work to be properly supervised by a properly qualified professional man such as an architect or surveyor...

repair and the occupier acted reasonably in entrusting the work to an independent contractor and had taken such steps (if any) as he reasonably ought in order to satisfy himself that the contractor was competent and that the work had been done properly. Judicial interpretation of the Act may now require a degree of supervision to be arranged by a client, who is deemed to be an occupier, as part of the steps that he reasonably ought to take. A contractor who has possession of a site will also be an occupier for the purposes of the Acts.

3.9 By virtue of the Unfair Contract Terms Act 1977, civil liability for death or personal injury due to negligence cannot be excluded or limited. Liability for property damage due to negligence can only be excluded or limited to the extent that the exclusion or limitation satisfies the statutory test of reasonableness.

3.10 Some insurance cover is expressly required by statute, e.g. the Employer's Liability (Compulsory Insurance) Act 1969. In addition, some insurance may be effectively necessitated by statute. For example, the Unfair Contract Terms Act 1977 includes the availability of insurance as a key consideration in the test for reasonableness of an exclusion or limitation clause. An individual or organisation may also be subject to specific insurance requirements, imposed through the contract covering the work or through the membership rules of a professional body, e.g. the Association of Consulting Engineers.

3.11 In the event that injury, loss or property damage is suffered on a project in which an organisation has been involved, the possibility exists of a claim being brought against the organisation, irrespective of whether a court will eventually hold the organisation liable. The conduct of the defence in such event will normally be subject to control by the organisation's insurers. It is essential to notify insurers promptly both of the occurrence of events which may give rise to claims, and of the receipt of any notices of claim. If such an accident has occurred, it is advisable for the organisation to investigate and record the facts carefully and ensure that there is a correct diagnosis of the cause.

3.12   Claims may also arise a long time after the relevant events, particularly in relation to damage to health, where the damage is not apparent immediately. Accordingly, records need to be retained.

3.13   Any organisation should recognise, in anticipation of the possibility of claims relating to both accidents and health, the need to be able to show, by means of reliable contemporary records, that the organisation had given thought to the relevant risks, had planned reasonable steps, and had actually implemented those steps. The steps might include training, supervision and the giving of appropriate warnings of risks and the precautions to be taken. The benefits of doing this are not just that it may provide a response to claims, nor just that it may also provide a defence to criminal charges under health and safety legislation as described below. The key benefit is that such forethought and precautions should also help to avoid accidents and damage to health and thereby minimise the likelihood of claims or prosecutions.

4   Statutory framework

4.1   The foundation of the statutory framework governing duties and enforcement powers relating to health and safety is now provided by the Health and Safety at Work etc. Act 1974. The raft of health and safety regulations currently in force are mostly made under powers conferred by the Act. Relevant parts of the Act and health and safety regulations are often referred to in regulations by the composite phrase 'the relevant statutory provisions'. (Previously 'building operations and works of engineering construction' were regulated by the Factories Acts and their subordinate 'Construction Regulations'. There was also other applicable legislation in related industries where construction professionals could be involved, such as the Mines and Quarries Act, now mostly repealed and replaced by health and safety regulations under the HSWA.)

4.2   The HSWA itself was born out of the Robens Report (1972), which recommended a shift away from the regime of detailed prescriptive regulations that existed then, on the grounds that they were unintelligible, inaccessible and ineffective. Robens recommended that health and safety law should be unified within a framework of a single comprehensive enactment, and that it should be goal-setting and aim to promote self-regulation rather than impose prescriptive rules.

4.3   Robens recognised that criminal sanctions were of limited value in preventing accidents, since prosecutions were always concerned with events that had already happened. He recognised the need for a constructive means of ensuring that practical improvements are made and preventive measures adopted. He recommended a constructive approach, but backed up with the availability of significant penalties to punish and deter those who were indifferent to their responsibilities for health and safety.

4.4   The HSWA carried most of the Robens' recommendations into effect. The HSWA laid down a number of general duties, and established the Health and Safety Commission (HSC) and the Health and Safety Executive (HSE) with special responsibilities and enforcement powers. The HSWA also conferred powers on the Secretary of State to make 'health and safety regulations', and on the HSC to issue or approve Approved Codes of Practice (ACoPs) to provide practical guidance on the interpretation of the general duties and health and safety regulations.

4.5   Subsequently, the power under the HSWA to make health and safety regulations has also been used as a vehicle for implementing EU Directives relating to health and safety, including, principally, the Health and Safety Framework Directive (1989) and the Temporary or Mobile Construction Sites Directive (1992). In general, the philosophy and approach of the EU Directives are compatible with the recommendations of the Robens Report.

4.6   The HSWA itself is occasionally subject to modification or amendment. Health and safety regulations and ACoPs are issued from time to time, and may be amended or replaced. Therefore, it is important to refer to a source of statutory materials which incorporates current amendments to the HSWA, and provides details and the current text of the health and safety regulations and ACoP currently in force. If dealing with an accident it is, of course, important to refer to the regulations, etc., that were in force at the relevant time, i.e. when the duty was to be exercised.

4.7   In addition, both the Act and regulations are subject to interpretation by the courts. The interpretations by the courts may give new meaning to the words of the statute, so construction professionals need to be aware of the key cases and their impact.

4.8   The HSWA and its subordinate regulations are enforced by the HSE or, in some circumstances, by local authority environmental health officers. The Act provides for enforcement not only through prosecution and penalties, but also through 'improvement' and 'prohibition' notices, which may be issued directly by an HSE Inspector. There is a statutory right of appeal to an Industrial Tribunal (and from there to the Employment Appeal Tribunal) to challenge such notices.

4.9   Section 33 of the HSWA creates the offences under the Act, including breach of the general duties and breach of health and safety regulations. It also lays down the possible modes of trial and the penalties. It provides for summary trial in a magistrates' court or, for more serious offences, trial on indictment in the Crown Court. The maximum penalties set originally by the HSWA for conviction on summary trial were gradually increased in line with inflation, but were then increased in 1992 by an order of magnitude, as recommended by the Cullen Report following the Piper Alpha disaster. There is no limit fixed in the Act on the fines that may be imposed in the Crown Court, but the level of

HEALTH AND SAFETY IN CONSTRUCTION

Case Study 5

In *R* v *Rollco* (1999), E, who was a specialist asbestos worker acting on this occasion as an independent contractor, was contracted to remove asbestos. He did not possess a licence. A vehicle hired by E was seen unloading bags of asbestos at a number of unauthorised sites — 200 or 300 bags had been fly-tipped on eight sites around Birmingham. He was sentenced to nine months' imprisonment.

fines actually imposed has increased similarly by an order of magnitude, in line with guidance from the Court of Criminal Appeal. There is limited provision for imprisonment, principally in cases of non-compliance with Prohibition Notices and in some other specific situations, such as contravening a relevant statutory provision by doing something without a required licence (which is particularly relevant to asbestos-related duties).

4.10   An individual may have duties and be liable to penalties under Section 33 as an employer, as a self-employed person or as an employee. In addition, Section 37 provides that where an offence has been committed by a body corporate (such as a limited company or local authority) and that offence is proved to have been committed with the consent or connivance of, or to have been attributable to any neglect on the part of an individual who was (or purported to act as) a director or manager or similar officer, the individual shall also be guilty of the offence and be liable to be proceeded against and punished accordingly.

4.11   The Court of Criminal Appeal has given guidance, in *R* v
        Howe (1999), that penalties should be fixed to reflect two
        elements: the gravity of the offence and the means of the
        offender. The gravity of the offence includes the factor 'how
        far short of the appropriate standard the defendant fell in
        failing to meet the reasonably practicable test'. Aggravating
        factors affecting the gravity of the offence include any loss of
        life, failure to heed warnings and whether the defendant
        intended to profit financially from the breach. Construction
        professionals will be aware from press reports of the level of
        fines imposed in high profile cases.

Case Study 6

In *R* v *Cardiff City Transport Services* (2001), a bus
company was convicted of breach of Section 2 of the
HSWA after an accident in which an employee driver
ran off the bus which he had just parked, and was killed
by another bus. The company had failed to carry out a
specific risk assessment relating to pedestrian and
vehicle movements. The Court of Criminal Appeal
accepted, however, that it was inappropriate to link the
death to the company's failing to carry out the risk
assessment, and reduced the fine imposed from
£75,000 to £40,000.

4.12   In addition to any fine, a convicted defendant will also
        normally be ordered to pay part or all of the prosecution
        costs, including the costs of any investigation by the HSE.

Case Study 7

In *R* v *Associated Octel* (1997), the Court of Appeal, dealing with submissions on the costs order which should be made against a guilty defendant, held that the costs of the prosecution which could be ordered to be paid by a defendant under Section 18(1) of the Prosecution of Offences Act 1985 may include the costs of the prosecuting authority in carrying out investigations with a view to the prosecution of a defendant where the defendant is convicted. (Such costs amounted in this case to over £100,000.) The court accepted that the prosecution could recover the whole of the HSE costs, as the defendant had not challenged the costs on the grounds that they were excessive or should not have been incurred, but the court recommended procedures for relevant information to be provided by the HSE in future and to allow any such challenge to be considered by the judge.

These costs can amount to very large sums, even if the defendant pleads guilty.

4.13   Fines cannot, in law, be covered by insurance, nor can they be recovered from another party to a contract. Insurance policies can, however, provide cover against the legal costs of defending a prosecution. There does not appear to be any public policy reason precluding insurance from covering orders to pay prosecution costs. Individual insurance policies should be checked. All insurance details need to be arranged when effecting the policy — it is too late after the event has occurred.

4.14   An unusual feature of the HSWA is that, by Section 40, the normal burden of proof is reversed. On a charge involving failure to do something 'so far as is reasonably practicable', for example, it is for the accused to prove that it was not reasonably practicable to do more than was in fact done to satisfy the duty or requirement. This makes defence very difficult if a prosecution is brought after an accident has occurred since, with the benefit of hindsight, it is usually possible to think of some step that might have avoided the accident. This tends to create a dilemma for construction professionals. Progress in health and safety demands that people should think what improvements could and should be made, as part of the process of learning from accidents. They should not be deterred from making such improvements. The courts recognise steps taken to remedy deficiencies after they are drawn to the defendant's attention as a mitigating factor in relation to sentencing, but those bringing prosecutions are inclined to argue that improvements made constitute an admission of earlier fault. Such arguments should be discouraged.

Case Study 8

After a railway accident, signalling arrangements were alleged to have caused an accident, but railway operators held back from making improvements, fearing that to do so would amount to an admission of responsibility for the accident.

4.15   Recent legislative developments have placed increasing emphasis on formal risk assessment, both as a specific mandatory requirement and as the basis of deciding what it is reasonably practicable to do to ensure the health and safety of employees and others. For guidance on the HSE's thinking on the risk assessment process, see *Reducing risks protecting people: HSE's decision making process* (2001).

5 HSWA general duties

5.1 The Health and Safety at Work etc. Act 1974 (HSWA) imposes general duties on employers, self-employed people, employees, those in control of premises (which include construction sites) and those who manufacture or supply items for use at work. The term 'employer' is defined in the EU Framework Directive as:

> *any natural or legal person who has an employment relationship with the worker and has responsibility for the undertaking and/or establishment.*

The term has been judicially explained in the context of UK health and safety law as meaning simply a legal person, including a company or public authority, who employs people. The HSWA imposes duties on employers which extend beyond the employer–employee relationship.

5.2 The HSWA is concerned with the health, safety and welfare of persons at work and the protection of other people from risks arising out of work activities. The 'general duties' on employers are contained in Sections 2, 3, 4 and 6. (Section 5 has been repealed and now forms part of environmental law.)

5.3   Section 2 concerns the duty of an employer towards his own employees. Section 2(1) states:

> *It shall be the duty of every employer to ensure, so far as is reasonably practicable, the health, safety and welfare at work of all his employees.*

Section 2(2) then sets out a non-exclusive list of matters to which the duty extends, which may be summarised as:

- provision of safe plant and systems
- use, handling, storage and transport of articles and substances
- provision of necessary information, instruction, training and supervision
- maintenance of the place of work to be safe and provision of safe means of access thereto and egress therefrom
- provision of a working environment which is safe and healthy and which includes adequate welfare arrangements.

5.4   Employers with five or more employees must have a written policy on health and safety which must include details of the organisation and arrangements in force for carrying out that policy.

5.5   Section 3 sets a duty on an employer (or self-employed person) to protect persons other than his own employees. Section 3(1) states:

> *It shall be the duty of every employer to conduct his undertaking in such a way as to ensure, so far as is reasonably practicable, that persons not in his*

employment who may be affected thereby are not thereby exposed to risks to their health or safety.

Section 3(2) imposes a similar duty on the self-employed. Section 3 does not contain any list of the matters to which the duty extends. Both employers and self-employed persons may be required, in prescribed circumstances, to provide information about the way they propose to conduct their work to others whose health and safety may be affected thereby.

5.6 Section 3 was enacted with the principal object of protecting the public from danger due to work operations, or workers employed by one employer being injured by the work operations of another employer, but it has been interpreted by the courts to create far-reaching indirect responsibilities, in two ways. Firstly, the courts have held that Section 3 imposes non-delegable duties on an employer, which transcend or override contractual boundaries or contractual allocation of risk or responsibilities. For example, employers have been held to be under a duty to provide information, warnings and safety

Case Study 9

In *R* v *Associated Octel* (1996), the House of Lords held that the 'undertaking' of the petrochemical company, Associated Octel, included repair work integrated with the general conduct of its business. C, an employee of a specialist contractor, was working on the repair of the lining of a tank using acetone (which is highly

continued overleaf

Case Study 9 continued

flammable) to clean the lining before applying fibreglass. He was using an ordinary lamp. The light bulb broke, causing a flash fire of the acetone vapour. Associated Octel was convicted of breach of Section 3 of the HSWA for failure to provide, or ensure that the employee was provided with, specialist equipment for working with acetone in a confined space. Lord Hoffmann stated:

Section 3 is not concerned with vicarious liability. It imposes a duty upon the employer himself. That duty is defined by reference to a certain kind of activity, namely the conduct by the employer of his undertaking. It is indifferent to the nature of the contractual relationship by which the employer chose to conduct it...If, therefore, the employer engages an independent contractor to do work which forms part of the conduct of the employer's undertaking, he must stipulate for whatever conditions are needed to avoid those risks and are reasonably practicable.

equipment to the employees of their contractors and sub-contractors, and for ensuring that their contractors and sub-contractors operate safe systems of work. It is no answer to say that the power to control was not contractually reserved — the duty requires that adequate powers of control should have been stipulated.

5.7   Secondly, the courts have extended the ambit of Section 3 to require clients, designers, contractors and verifiers during

> ## Case Study 10
>
> Following the Ramsgate walkway collapse, in which a number of members of the public were killed, the client port authority, the design and construct contractor, the contractor's designer, and the third party certification body, were all convicted of breaches of the HSWA general duty under Section 3 on the basis that they had failed to do all that was reasonably necessary during the planning, design, construction and verification to safeguard against **risks to the public after completion** arising from defective design and construction, or from risks inherent in the design.

design, construction and verification phases to have regard not only to safety during the construction phase, but also safety of the completed works in use.

5.8 Section 4 imposes a duty on *'each person who has, to any extent, control of non-domestic premises'* which have been made available to persons who are not their employees *'as a place of work or as a place where they may use plant or substances provided for their use there'*. The duty is for the person having control of the premises to:

> *take such measures as it is reasonable for a person in his position to take to ensure so far as possible that the premises, access thereto and egress therefrom available for use by persons using the premises and any plant or substances in the premises or provided for use there are safe and without risks to health.*

the most useful purchase as a source of regulatory requirements. They usually incorporate the text of the related regulations. Some ACoPs also include some non-ACoP material, which is for guidance only and does not have the quasi-statutory status of an ACoP. The respective texts are sometimes distinguished by the printing format, e.g. the regulations in italics, the ACoP material in bold type, and the non-ACoP guidance in plain type. Some of the earlier regulations have no ACoP. There are no ACoPs directly relating to the HSWA general duties or the Construction (Health, Safety and Welfare) Regulations 1996.

6.3   Other guidance documents are published by the HSE, e.g. *Health and Safety in Construction* (2001), but these documents do not have any special legal status and, except where such material is included as part of an ACoP, there is usually no identification of the regulations on which statements are based. Note that documents published by the HSE commonly do not include reference to the requirements

Case Study 11

In order to minimise health and safety risks due to exposure of people at work to handling cement while they were installing ground anchors into gravel strata, the designer decided to specify a chemical grout delivered in liquid form direct to a tank ready for injection. However, pollutants from the chemical grout leached into the gravel aquifer. This created an offence under environmental legislation.

of related regulations which are not classified as health and safety regulations, e.g. environmental regulations on contaminated ground or on waste management dealing with asbestos. Construction professionals need to consider all relevant risks and regulations.

6.4 Health and safety regulations can be classified under various categories, as set out below, with some of the more significant regulations in each category listed.

General regulations concerning health and safety management

- Management of Health and Safety at Work Regulations 1999 (MHSWR) (replacing MHSWR 1992)
- Construction (Design and Management) Regulations 1994 (CDM) (amended 2000 — also new ACoP 2001 in force 2002)
- Reporting of Injuries, Diseases and Dangerous Occurrences Regulations 1995 (RIDDOR)
- Control of Major Accident Hazards Regulations 1999 (COMAH)
- Personal Protective Equipment Regulations 1992/99 (PPE)
- Health and Safety (First Aid) Regulations 1981

Regulations concerning workplaces and dangerous substances

- Workplace (Health, Safety and Welfare) Regulations 1992
- Construction (Health, Safety and Welfare) Regulations 1996
- Control of Substances Hazardous to Health Regulations 1999 (COSHH)

- Control of Asbestos at Work Regulations 1987
- Control of Lead at Work Regulations 1998
- Fire Precautions (Workplace) Regulations 1997 (amended 1999)
- Noise at Work Regulations 1989

Regulations concerning machinery, handling and equipment

- Lifting Operations and Lifting Equipment Regulations 1998 (LOLER)
- Provision and Use of Work Equipment Regulations 1998 (PUWER)
- Supply of Machinery (Safety) Regulations 1992
- Manual Handling Operations Regulations 1992

Regulations relating to specific areas of work or situations

- Docks Regulations 1988
- Railways (Safety Case) Regulations 1994
- Quarries Miscellaneous Health and Safety Provisions Regulations 1995
- Pipelines Safety Regulations 1996
- Confined Spaces Regulations 1997
- Work in Compressed Air Regulations 1996
- Diving at Work Regulations 1997

6.5   The regulations are not structured to be mutually exclusive. It is a major task to ensure that all relevant regulations have been identified, to understand their meaning and to keep abreast of changes in regulations. The difficulty of doing so provides no excuse in law for failure to do so.

Case Study 12

CIRIA Report C518 *Safety in ports — ship-to-shore linkspans and walkways,* commissioned after the Ramsgate walkway failure, identified seven sets of regulations as particularly relevant to the procurement, operation and maintenance of ship-to-shore linkspans and walkways. The document provides guidance to the requirements of the regulations. It demonstrates how accessibility and understanding of the regulations can be improved, particularly in the context of specific types of work or projects, through construction industry co-operation.

6.6   The Workplace (Health, Safety and Welfare) Regulations 1992 and the Fire Precautions (Workplace) Regulations 1997 do not generally apply to construction sites (although they may apply to related workplaces such as casting yards). Such matters are covered on construction sites by the Construction (Health, Safety and Welfare) Regulations 1996. However, both the Workplace and Fire Precautions Regulations apply to other workplaces and may have direct or indirect relevance. For example, the Workplace Regulations contain an important provision in Regulation 5 regarding the duty of an employer to maintain workplace structures — this is important to construction professionals, since they may be involved in inspection and maintenance work. Also designers must consider generally whether the works they design will comply with Regulations such as the Workplace Regulations.

6.7   There is an issue as to how the duties under health and safety regulations, made under HSWA Section 15, fit with the HSWA general duties. Does the breach of a regulation constitute evidence of breach of the general duties? More importantly, does compliance with regulations constitute evidence of satisfying the general duties? The answer may depend on whether the relevant regulations are implementing an EU Directive, since UK safety legislation should be interpreted generally to be consistent with EU Safety Directives. It appears to be a sound principle (although the courts may yet decide otherwise) that the framework of responsibilities laid down by the Construction (Design and Management) Regulations 1994 indicates who is deemed to be responsible for what on construction projects.

6.8   Some of the specific regulations are of particular significance. For example, asbestos is one of the most pernicious causes of premature death, and the HSE is keen to prosecute failures to comply with the regulations relating to asbestos. However, for the purpose of this general introduction, the three sets of health and safety regulations of most significance are the Management of Health and Safety at Work Regulations 1999 (MHSWR), the Construction (Design and Management) Regulations 1994 (CDM) and the Construction (Health, Safety and Welfare) Regulations 1996 (CHSWR). A brief outline of each is given below.

7 MHSWR and risk assessment

7.1 The Management of Health and Safety at Work Regulations 1999 (MHSWR) (which replace the Management of Health and Safety at Work Regulations 1992) are not construction specific. They implement the EU Health and Safety Framework Directive (1989) and cover:

- risk assessment and the principles of prevention
- health and safety arrangements
- health and safety personnel and advisers
- procedures and arrangements for emergencies
- information for employees
- co-operation and co-ordination
- employees' capabilities and training
- employees' duties.

7.2 Reference should be made to the full text of the Regulations, ACoP and non-ACoP guidance. It has been suggested (J. Anderson, *ICE Proceedings* 2001) that the MHSWR are so fundamental that they ought really to be incorporated as part of the primary legislation. There is some overlap with the HSWA general duties.

7.3 As required by the EU Framework Directive, Regulation 3 of the MHSWR makes formal assessment of risks to health and safety a mandatory requirement. Every employer and self-employed person is required to assess the risks both to his own employees or himself while at work, and to *'persons not in his employment arising out of or in connection with the conduct by him of his undertaking'*. The assessment is to be *'suitable and sufficient'*. Guidance on what is suitable and sufficient relative to different situations is given in the ACoP, which states that the level of risk arising from the work activity should determine the degree of sophistication of the risk assessment.

7.4   Regulation 3(3) requires assessments to be reviewed as and when circumstances change. Regulation 3(5) imposes special risk assessment requirements relating to employees who are young persons. Regulation 3(6) requires that where there are five or more employees, the significant findings of the risk assessment must be recorded.

7.5   Formal risk assessment involves several steps, starting with identification of both sources and targets of risk. The MHSWR ACoP paragraph 11 defines the terms 'hazard' and 'risk' as follows:

    (a)   *a hazard is something with the potential to cause harm (this can include articles, substances, plant or machines, methods of work, the working environment and other aspects of work organisation);*

    (b)   *a risk is the likelihood of potential harm from that hazard being realised. The extent of the risk will depend on:*

        (i)   *the likelihood of that harm occurring;*

> *(ii) the potential severity of that harm, i.e. of any resultant injury or adverse health effect; and*
>
> *(iii) the population which might be affected by the hazard, i.e. the number of people who might be exposed.*

7.6 These definitions are useful as a basis for evaluating risks, but engineers should not be overawed by the terminology. Some sources of risk may be naturally described as 'hazards', e.g. underground electric cables, but control of the risk may require closer focus on specific mechanisms or event trails that might lead to harm. For example, there is the possibility that information on the location of underground electric cables might be unreliable, or that a cable might not be detected by the available equipment, or by the personnel using the equipment. These are sources of risk, but they might not be considered as 'hazards'.

7.7 Some techniques have been developed to assist risk assessment, such as the creation of a 'risk register'. Such techniques may be helpful, but should not be allowed to dominate thinking, for example, by pressurising the construction professional to fit the risk description into words in a pro forma box of a certain size. Construction professionals may well find drawings and diagrams a more powerful tool than lists to identify and communicate the location and nature of risks.

7.8 The scope of the requirement for every employer and self-employed person to consider risks to persons not in his employment extends not only to the public who might be affected by the undertaking, but also to the employees of co-contractors, sub-contractors and all those working on the same project. It is not limited to risks while the activity is in progress. It includes the effect of design or construction work on future users of the works or facility.

7.9   The purpose of the risk assessment is to enable the employer or self-employed person to identify the measures they need to take, as part of their undertaking, to comply with *'the requirements and provisions imposed upon him by or under the relevant statutory provisions and by Part II of the Fire Precautions (Workplace) Regulations 1997'*. The relevant statutory provisions comprise the HSWA general duties and any duties under health and safety regulations. A 'suitable and sufficient risk assessment' should identify, in a structured manner, the significant risks arising out of work, to enable the employer or the self-employed person to identify and prioritise the measures that are needed to comply with the law. These measures should be appropriate to the nature of the work being undertaken. According to the ACoP guidance, the assessment should record the preventive and protective measures in place to control the risks identified and, insofar as existing measures are not adequate, consider what more should be done to reduce risk sufficiently. The ACoP states that:

> *once the risks are assessed and taken into account, insignificant risks can usually be ignored, as can risks arising from routine activities associated with life in general, unless the work activity compounds or significantly alters those risks.*

7.10   The ACoP recognises that specific risk assessments may be required by other regulations of general application, e.g. COSHH or the Manual Handling Regulations, or that special levels of risk assessment may be laid down by regulations covering special situations, e.g. COMAH or the Railway (Safety Case) Regulations. The MHSWR ACoP paragraphs 27/28 provide that a further risk assessment is not required to satisfy the MHSWR.

7.11   The level of risk may well depend on the individuals or organi-sations involved in carrying out the work activity — whether the employees are particularly skilled and experienced or, at the other extreme, have some disability. There is accordingly a link with MHSWR Regulation 13, which requires that every employer shall, in entrusting tasks to his employees, take into account their capabilities as regards health and safety.

7.12   MHSWR Regulation 4 states:

> *Where an employer implements any preventive and protective measures he shall do so on the basis of the principles specified in Schedule 1 to the Regulations.*

Schedule 1 repeats the 'principles of prevention' laid down in Article 6 of the EU Framework Directive. These lay down a hierarchical approach to prevention and protective measures. The first principle is stated as 'avoiding risks'. While compliance with the principles of prevention is a mandatory requirement, their application is a matter of legal interpretation. The courts may provide a ruling if a case arises. Excessive literal interpretation is undesirable, particularly at the level of individual risks. If the words were taken too literally, construction professionals would never build anything. There is a particular danger that some risks might be avoided at the expense of creating or increasing other risks. Construction management and design is about dealing with the totality of risk, as well as individual risks. It is about facing up to and dealing with risks to provide an acceptable level of safety.

7.13   Rules on the interpretation of statutory regulations derived from EU Directives allow regard to the purpose of the Directive and Regulations. The overarching purpose of the Framework Directive is stated in Article 5(1), that *the employer shall have a duty to ensure the safety and health of workers in every aspect related to the work'*. This supports the view that regard should be had to the totality of risk as well as to individual risks.

7.14   In construction, the principle of avoiding risks requires, as a matter of high level policy, a general move away from specific risks identified as avoidable, for example:

- replacing work from ladders by the use of general access scaffolds, tower scaffolds, mobile elevating work platforms (MEWPs) or similar for working at height
- replacing solvent-based substances with non-solvent based ones
- using lower voltage electrical equipment in place of high voltage
- supplying materials in units weighing not more than 25 kg for manual lifting by individuals.

7.15   This policy is best implemented through industry-wide action, since it depends on the development and availability of relevant equipment, packaging, materials and other products, particularly ensuring that any substitute materials or methods do not introduce different risks to health and safety. The principle does not provide any defence if, for example, the substitute adhesives allow tiles to fall off, or the substitute paints reduce durability. Construction professionals

Case Study 13

Large-diameter augered piles might be marginally the best foundation engineering solution for a large structure to be built on contaminated land, but the alternative of driven piles would eliminate the process bringing hazardous materials to the surface and thereby exposing workers and others to foreseeable risks. The alternative might be more expensive. It might create other health risks such as noise. The designer would need to consider and weigh all these matters.

need to adopt a composite, integrated approach to safety. Application of the principle at a project-specific level can be illustrated by an example (see Case Study 13).

7.16   The remaining principles of protection and prevention follow after 'evaluating the risks which cannot be avoided'. They include:

- *Reduction.* There are some areas where it is not possible to eliminate risks without compromising other essential features of a project, but if it is possible to reduce them this must be done.
- *Collective protection.* Collective protection which does not rely on discipline or spasmodic action by individuals for its effectiveness should be provided. For example, acoustic shielding of a noisy compressor or — better still — its replacement by a less noisy machine should be used wherever possible in preference to hearing protection being provided.

- *Organisational controls.* Organisational controls are special rules or procedures. The permit to work system is the most common. Such procedures necessarily rely for their effectiveness on someone taking specific action and the possible failure by an individual to take the specific action should be recognised as a risk.
- *Personal protection.* Personal protection should not be relied upon as the sole or primary method of risk control if there are other methods of providing protection and prevention, but personal protection must still be provided, for example, to comply with the Construction (Head Protection) Regulations 1989 and the Personal Protective Equipment Regulations 1999, where an identified risk exists.

7.17 The Robens and Cullen Reports both reacted against a prescriptive regime in favour of 'goal-setting' for the development of safety procedures and practices. An illustration of a goal-setting approach is included in Appendix C. However, construction professionals may nevertheless find it effective to deal with common or generic risks by a code of practice of prescriptive requirements. This may help to establish a consensus on what steps are appropriate to reduce common risks sufficiently and what levels of risk are acceptable. Research suggests that the number of risks that can be actively managed at any one time is limited. Common risks should be addressed through the development of safe habits and practices, instilled by education, example and supervision, rather than reliance on bulky documents, which are liable to be unread and whose bulk may obscure the focus on specific risks.

7.18 MHSWR Regulation 5 requires every employer to make and give effect to such arrangements as are appropriate, having regard to the nature of his activities and the size of his undertaking, for the effective planning, organisation, control, monitoring and review of the preventive and protective measures identified as necessary by a risk assessment, and to record the arrangements if five or more people are employed. This may be regarded as either overlapping or elaborating the requirements of HSWA Sections 2(3) and (4).

8 CDM Regulations

8.1   The Construction (Design and Management) Regulations 1994 (as amended 2000) (CDM) implement most of the EU Temporary and Mobile Construction Sites Directive (1992). A new ACoP was issued in 2001 (in force 2002). The CDM Regulations are specific to the management of health and safety at work in relation to projects which include or are intended to include 'construction work' as defined. The definition includes site clearance, investigation and excavation, maintenance, refurbishment, alteration and demolition work. The CDM Regulations contain exemptions for certain minor works, but demolition work of any nature is never exempted, and the requirements on designers apply even to exempted works.

8.2   The philosophy of the CDM Regulations stems from the realisation and acknowledgement that (as good construction professionals have always known) the seeds of risk and safety are sown at all stages of a project, starting at its inception long before the construction work commences on site, and that effects on safety carry through into cleaning and maintenance work during the life of the structure. In particular, the CDM Regulations recognise that:

Case Study 14

The client for a large new commercial development, which involved extensive cut and fill operations as site preparation, identified at an early stage the risks involved in deep excavations for drainage. Accordingly, the client took steps to ensure that both the design and the planning of the construction sequence were carried out with a view to minimising the depth of excavation required while ensuring that the drainage would operate satisfactorily on completion (e.g. by installing pipe runs before placing fill, but ensuring that the pipes could accommodate settlement).

- information relevant to safety needs to be gathered and made available for the construction work, and for subsequent cleaning, maintenance and demolition work
- permanent works designers should consider the risks to safety during the construction work and subsequent cleaning, maintenance and demolition work, which may flow from their design decisions. Temporary works designers should consider risks to safety during the construction work for which the temporary works are provided and during any use, maintenance and demolition or removal of the temporary works.

8.3 The CDM Regulations place specific duties on clients, designers and contractors to consider health and safety so that it is taken into account and then co-ordinated and managed effectively throughout all stages of a construction project — from conception, design and planning, through to the execution of works on site, subsequent maintenance and repair, and eventual demolition. Two new roles, the planning supervisor and principal contractor, have been introduced to provide health and safety co-ordination and perform certain health and safety functions.

8.4   For many of the duty holders, these regulations involve a radical change in culture. It was, in earlier times, allowed that clients should not be expected to make a contribution to the control of health and safety risks beyond appointing apparently competent contractors and designers. Designers of the permanent works were not required by law to consider the implications of their designs for the safety of persons constructing the works.

8.5   Under the CDM Regulations, a client has duties to:

- provide information about the premises where construction work is intended — this includes information about site conditions or the conditions of existing structures, and may require investigation to be carried out to obtain the information
- ensure the competence of the planning supervisor, designers and contractors who the client intends to appoint, and the adequacy of their resources allocated to carry out their duties under the CDM Regulations
- have produced a pre-tender health and safety plan containing relevant information and to give this to contractors at tender stage
- ensure that a construction phase health and safety plan complying with Regulation 15(4) has been produced before the construction phase starts (breach of this duty can give rise to civil liability)

Case Study 15

An existing retaining wall at the foot of a slope showed signs of distress. The problem was diagnosed as inadequate reinforcement in the existing wall. A decision was made by the body responsible to reconstruct with an enlarged wall. Partly because of the small scale of the project, no detailed site investigation was carried out. A contract was let without any detailed assessment of the risks associated with construction. The designer assumed that it would be possible to construct the works safely as designed without substantial temporary works. The soil conditions turned out to be worse than assumed. A major slip occurred when the contractor started excavation and demolition of the wall, not expecting such conditions. Fortunately nobody was killed. The works had to be completely redesigned to enable safe construction.

- appoint a planning supervisor to ensure the production of a health and safety file, which will provide safety related information for those involved on subsequent cleaning, maintenance and demolition work, and to ensure that the information in the health and safety file received is kept available for inspection by any person who may need it for relevant purposes.

8.6   The scheme of the CDM Regulations for the pre-construction phase of the project involves the appointment by the client of a person to act as 'planning supervisor', who then has duties and responsibilities (but very little by way of direct powers) laid down by the Regulations. The client may appoint a member of his own staff or a department within his own organisation to carry out the role of planning supervisor, but the client is subject to a general requirement to ensure that the person appointed is competent and will allocate adequate resources to perform the functions of planning supervisor. The appointment may be transferred to a different person during the project. One of the designers can be appointed as planning supervisor. The primary role of the planning supervisor during the pre-construction phase is essentially to advise the client, ensure that the designers discharge their duties under the CDM Regulations, and co-ordinate the client's and designers' outputs to produce the pre-tender health and safety plan.

8.7   The planning supervisor also has a role of assessing the competence and allocated resources of prospective contractors and designers at all levels, so as to be in a position to advise the client appointing a contractor or designer, and any contractor appointing a designer, on the sufficiency of such competence and allocated resources. The client and any contractor appointing a designer must be satisfied that the person to be appointed has the competence to prepare the design and has allocated, or will allocate, adequate resources to comply with their duties under the CDM Regulations. The client or any contractor appointing a contractor must be satisfied that the person to be appointed has the competence and has allocated, or will allocate, adequate resources *'to perform any requirement and to conduct his undertaking without contravening any prohibition, imposed on him by or under any of the relevant statutory provisions'*. The 'relevant statutory provisions' referred to in the criteria for competence and resources, extend beyond the CDM Regulations. They include the HSWA general duties and other health and safety regulations. The scope of the competence and resources to be considered is therefore very wide. There needs to be a structured and meaningful procedure to consider whether contractors and designers to be appointed are competent and have allocated, or will allocate, adequate resources. There is no requirement, however, for the procedure to be started afresh for each project — it should be possible to streamline prequalification processes, for example, by establishing and using registers of designers and contractors.

8.8 Designers are under a duty to assess and eliminate or control the health and safety risk implications of their designs for:

- persons at work performing the construction work
- persons cleaning windows or other transparent or translucent parts of a building
- possible persons at work on subsequent construction work, which is defined to include decoration, maintenance and demolition work
- any person who may be affected by the work of such persons at work.

8.9 The terms 'design' and 'designer' are defined in the CDM Regulations. The 2001 ACoP provides a gloss on the definition of 'designer' to extend it to anyone, including a client, who dictates or alters a design, or who specifies the use of a particular method of work or material. It is also stated to include a contractor purchasing materials where the choice has been left open. This extended definition has implications for the client or contractor in regard to ensuring the relevant competence of contractors to be appointed, and for planning supervisors who are co-ordinating 'designers', as well as for the contractors who have duties as designers.

Case Study 16

A large project was procured on a nominal design-and-construct basis, but the preliminary design had been produced by the client and the contract stipulated that the contractor's design (including any deviations from the preliminary design) would be subject to approval by the client. The client appointed an independent planning supervisor. The new planning supervisor was asked by the client, as an additional task, to audit the contractor's health and safety performance on site. This he did enthusiastically. However, the planning supervisor felt inhibited from questioning or criticising the client's actions as a designer, which included the client's actions in approving the contractor's design. The client refused, because of cost implications, to approve a proposal by the contractor's designer to change the preliminary design to reduce a risk during construction. The planning supervisor made no comment. A serious incident occurred during construction, which would have been avoided if the proposed change had been approved.

8.10   The designer's duties are, so far as is reasonably practicable, to:

- alert clients to their duties
- consider during the development of designs the hazards and risks which may arise to those constructing and maintaining the structure

- design to avoid foreseeable risks to health and safety
- eliminate risks to health and safety by design choices, if this is possible
- reduce risks at source if avoidance is not possible
- consider measures which will protect all workers if neither avoidance nor reduction to a safe level is possible
- ensure that the design includes adequate information on health and safety issues
- pass this information on to the planning supervisor so that it can be included in the health and safety plan, and ensure that it is given on drawings or in specifications and elsewhere as appropriate
- co-operate with the planning supervisor and, where necessary, other designers involved in the project.

8.11 The duties of designers as described under the CDM Regulations effectively entail risk assessment at various stages of the design process (although the duties are not defined by the CDM Regulations or the related ACoP as a 'risk assessment') looking at:

- risks which could arise in the construction of the works that they are designing, in relation to the people at work on the construction site, and people affected by the activities of such people at work
- risks which could arise in subsequent cleaning of translucent panels, and during subsequent construction work (which is defined to include maintenance and demolition) in relation to the people at work on such cleaning or subsequent construction work, and people affected by the actions of such people at work.

This means that permanent and temporary works designers need to think through how their designs might be constructed, cleaned, maintained and demolished. On the other hand, it does not require that they specify how the designs shall be constructed, cleaned, maintained or demolished. Designers should also consider the effect of any changes proposed to improve health and safety during construction on the subsequent durability, performance and use of the permanent works. These longer-term effects may also be health and safety risks and are required to be considered under the MHSWR. Risks need to be considered in the round, as the avoidance of one risk may create or increase other risks. Any effects of demarcation between the CDM Regulations and the MHSWR need to be overcome.

8.12   For the construction phase, the client is required to appoint a person who is a contractor (as defined in the CDM Regulations, see below) to act as 'principal contractor' in respect of the project. The powers and duties of the principal contractor are laid down in the Regulations. In particular, the principal contractor is required to produce a construction phase health and safety plan, based on and responding to the pre-tender health and safety plan.

8.13   The principal contractor's key duties are to:

- co-ordinate the activities of all contractors so that they comply with health and safety law
- develop and implement the health and safety plan
- where the work is sub-contracted, arrange for competent and adequately resourced contractors to carry it out
- ensure the co-ordination and co-operation of contractors
- obtain from contractors the main findings of their risk assessments and details of how they intend to control and manage the risks

- ensure that contractors have information about risks on site
- ensure that workers on site have been given required health and safety training
- ensure that contractors and workers comply with any site rules which may have been set out in the health and safety plan
- monitor health and safety performance
- ensure that all workers are properly informed and consulted
- make sure only authorised people are allowed on to the site
- pass information to the planning supervisor for the health and safety file.

8.14   The CDM Regulations confer special powers on the principal contractor, which override any contractual arrangements. The principal contractor **may**:

- give reasonable directions to any contractor on health and safety matters
- include in the construction phase health and safety plan, rules for the management of the work.

8.15   The planning supervisor has no responsibility under the CDM Regulations for the principal contractor, except:

- to advise the client before appointment of the principal contractor whether or not the proposed principal contractor has the competence to act as principal contractor and has allocated or will allocate adequate resources to comply with his responsibilities under the Regulations
- to advise the client whether or not the principal contractor has produced a construction phase health and

safety plan complying with Regulation 15(4) before the construction phase starts.

8.16   The principal contractor has a specific duty to take reasonable steps to exclude unauthorised persons from the premises or part of the premises where construction work is being carried out. Breach of this duty can give rise to civil, as well as criminal, liability.

8.17   The CDM Regulations have also introduced the requirement of a health and safety file, which will provide information for users, future designers and contractors, and those who will carry out cleaning and demolition work. The planning supervisor has the function of compiling the information (or ensuring that the information is compiled) for the health and safety file.

8.18   A contractor is defined to mean any person:

who carries on a trade, business or other undertaking (whether for profit or not) in connection with which he –

(a)   *undertakes to or does carry out or manage construction work,*

(b)   *arranges for any person at work under his control (including, where he is an employer, any employee of his) to carry out or manage construction work.*

8.19   Every contractor (which includes any sub-contractor) is under a duty to:

•   co-operate with the principal contractor

- provide information for the health and safety plan
- provide information about risks to health and safety arising from their work and steps they will take to control and manage the risks
- manage their work so that they comply with rules in the health and safety plan
- provide information about reportable injuries, dangerous occurrences and ill health
- provide information to their employees.

9   CHSWR

9.1   The Construction (Health, Safety and Welfare) Regulations 1996 (CHSWR) implemented Annex IV of the EU Temporary or Mobile Construction Sites Directive, and consolidated three sets of earlier construction regulations. They are based on the same broad definition of 'construction work' as in the CDM Regulations. They provide the general regulations relating to operations and welfare on construction sites. The CHSWR apply generally to, and in relation to, construction work carried out by a person at work.

9.2   The duties and requirements under CHSW Regulations are generally clear and self-explanatory. They deal with specific hazards, systems of work, competency and training. No additional guidance is required except, perhaps, to note particularly the requirement for specific inspections and reports in Regulations 29 and 30 and Schedules 7 and 8. What should be emphasised is how the Regulations apply to employers, self-employed persons, employees and those with supervisory functions.

9.3   The CHSW Regulations impose duties, firstly, on:

- every employer whose employees are carrying out construction work

- every self-employed person carrying out construction work.

9.4 The scope of their duty is *'to comply with the provisions of these Regulations insofar as they affect him or any person at work under his control or relate to matters which are within his control'*. Secondly, the Regulations impose a duty on every employee carrying out construction work *'to comply with the requirements of these Regulations insofar as they relate to the performance of or the refraining from any act by him'*. Thirdly, they impose a duty on any person who does not have a duty as employer, self-employed person or employee carrying out construction work, but who *'controls the way in which any construction work is carried out by a person at work'*. This would probably include a resident engineer or clerk of works, as well as construction professionals employed or appointed by the contractor. The power to control may have been stipulated in a contract, or it might arise by statute. The duty on such persons is *'to comply with the provisions of these Regulations insofar as they relate to matters which are within his control'*.

9.5 Finally, there is a general duty on every person at work to co-operate in regard to duties and requirements to be performed or complied with under the Regulations, and a general duty to warn. The duty to warn is stated as follows:

> *It shall be the duty of every person at work, where working under the control of another person, to report to that person any defect which he is aware may endanger the health or safety of himself or another person.*

As a matter of statutory interpretation, this duty to warn probably only extends to risks addressed by the CHSW Regulations, but see also Chapter 11 below on the general topic of the duty to warn.

10 Contractual provisions

10.1   Traditionally construction contracts were used to relieve clients of responsibility and liability for health and safety risks during construction, as regards both workmen and the public. This arrangement was accepted by the courts and parliament (prior to the 1980s or thereabouts) as effective in relation to both criminal and civil liability, subject to the client having acted reasonably in entrusting the work to the independent contractor and taken such steps (if any) as he reasonably ought to satisfy himself that the contractor was competent and that the work had been properly done.

10.2   This explains why the early editions of, for example, the *ICE Conditions of Contract* made little reference to health and safety during construction. The only relevant provisions in the first four editions of the *ICE Conditions of Contract* were:

- Clause 19, which required fencing and lighting to exclude and protect the public
- Clauses 22 and 24, which excluded the Employer from liability in respect of any accident or injury to any member of the public or to any workman, unless it resulted from any act or default of the Employer, his agents or servants

- Clause 23, which required the Contractor to indemnify the Employer, and to effect relevant insurance against third party claims.

(The requirements of Clauses 22, 23 and 24 have been retained in later editions, but they are aimed principally at insurance and are not considered further here.)

10.3　The Committee responsible for the Fifth Edition of the *ICE Conditions of Contract*, published in 1973, anticipated or responded to the message of the Robens Report (1972) by introducing significant new health and safety requirements and powers. The other main standard forms of construction contract of the time, GC/Works/1 and JCT, did not respond to the same extent, presumably considering it inappropriate to duplicate the statutory duties.

10.4　Contractual provisions can impose requirements in relation to health and safety, and provide a framework for supervising and enforcing the requirements. They do not and cannot override the statutory duties specified in the HSWA and health and safety regulations. They may, however, provide an effective means or framework for discharging those duties, particularly having regard to the judicial ruling that HSWA Section 3 requires clients to stipulate and exercise appropriate levels of control in regard to health and safety matters.

10.5　Within the latest editions of the *ICE Conditions of Contract*, relevant provisions have been made, including the following.

- Clause 8 was amended to set out a division of responsibility between Engineer and Contractor for permanent works and construction work (including temporary works). It provided that *'the Contractor shall take full responsibility for the adequacy stability and safety of all site operations and methods of construction'*. This now needs reconsideration in the light of the CDM Regulations.

- Clause 14 introduced extensive new powers for the Engineer to require submission by the Contractor of details of arrangements and methods of construction. It empowered the Engineer to withhold consent or impose special requirements if not satisfied.

- Clause 15, dealing with superintendence of the works, was extended to require that *'such superintendence shall be given by sufficient persons having adequate knowledge of the operations to be carried out (including the methods and techniques required for the hazards likely to be encountered and methods of preventing accidents)'* as may be requisite for the satisfactory and safe construction of the Works.

- Clause 16, dealing with the removal of incompetent employees, was extended to allow the Engineer to require the removal of any person on the grounds that he *'fails to conform with any particular provision with regard to safety which may be set out in the specification or persists in any conduct which is prejudicial to health and safety'*.

- Clause 19, watching and fencing, was extended to impose general duties for health and safety on the Contractor, and on the Employer as regards work carried out by his own workmen or by other contractors on the site.

- Clause 71 has been added to allow for the effect of the CDM Regulations, making the actions of the planning supervisor and principal contractor equivalent to an Engineer's instruction pursuant to Clause 13. This gives such actions contractual force, in addition to their statutory force, and deals with issues of consequential payment.

10.6 In the *NEC Engineering and Construction Contract* (NEC), there is provision for health and safety obligations to be stated as part of the Works Information. Clause 18.1 states:

> *The Contractor acts in accordance with the health and safety requirements in the Works Information.*

There is also provision for termination by the Employer, in some circumstances, following a substantial breach of health and safety regulations by the Contractor.

10.7 Other recent standard forms of construction contract, e.g. JCT 98 and GC/Works/1 (1998), make provision for the CDM Regulations but have still chosen not to provide detailed powers and duties separate from the powers and duties imposed by HSWA and health and safety regulations.

10.8 There is a practical difficulty in imposing health and safety obligations through contract, namely that the normal legal remedy for breach of contract, i.e. damages, is inadequate as a sanction. A breach of a term imposing health and safety requirements is unlikely to affect the final product, so there will be no loss caused, and no damages will be recoverable for breach of the term. Insofar as the Employer might suffer a fine arising out of an accident caused by a breach of health and safety obligations by the Contractor, the Employer would not be able to recover such fine from the Contractor. Accordingly, more direct powers, for example the power to withhold consents or the power to order individuals off the site in the event of a breach, must be stipulated.

10.9   Direct powers can be specifically reserved in the contract to suspend the progress of the works or to terminate the employment of the contractor in the event of breach of health and safety regulations, but such powers should be exercised cautiously. Under the *ICE Conditions of Contract*, Clauses 40 and 63 provide some relevant powers but they have not been designed for this purpose. Legal advice should be taken before exercising such powers. The NEC provides some express powers of termination relating to substantial breaches of health and safety regulations, based on notice and failure to stop defaulting. Again, legal advice should be sought at an early stage before exercising such powers. Suspension or termination without legal justification can be very expensive.

10.10   Contracts should be drafted to be compatible and consistent with the statutory duties and powers, and to acknowledge their implications. For example, statutory duties, particularly duties to provide information, may have commercial significance. Contractual provisions that are intended to shift or delegate responsibilities may not be effective as regards criminal liability, if the responsibility is non-delegable.

10.11   Statutory health and safety duties and powers also extend across contractual boundaries. For example, a main contractor has responsibilities for the safety of a sub-contractor's employees, a designer has responsibilities for risks to safety during construction arising from the design, a principal contractor has statutory powers not dependent on contract.

11 Duty to warn

11.1   A construction professional who, in the course of general or specific duties, notes any defect or failure to observe safety requirements, should first take such action as is immediately available and appropriate, informing the contractor or other body having responsibility. Initially this should be done at working level, but formal reporting at a higher level may be desirable to prevent repetition. This is in addition to any duty to warn imposed by the CHSW Regulations.

11.2   If the construction professional finds that the defect or failure is still not corrected, or otherwise encounters a potentially disastrous situation (i.e. one which could lead to death or serious injury or property damage) in circumstances which do not fall within established procedures, legal duties of care, coupled with Rules for Professional Conduct, require the construction professional to react responsibly to such a situation. Guidelines prepared by the Royal Academy of Engineering (formerly the Fellowship of Engineering) provide useful guidance in such circumstances. A copy is attached as Appendix B.

11.3   Workers who disclose information relating to health and safety matters now have statutory protection of employment rights in defined situations under the Employment Rights

Act 1996 Part IVA, as inserted by the Public Interest Disclosure Act 1998. 'Workers' are defined to include individuals working on an agency basis as well as employees working under a contract of employment. For the protection rights to apply, the disclosure must be a 'qualifying disclosure' as regards content, the person to whom it is made and the motivation for making it.

11.4   As regards content, the information disclosed must, in the reasonable belief of the worker making the disclosure, tend to show that a person has failed, or is failing or likely to fail to comply with any legal obligation to which he is subject, or that the health or safety of any individual has been, is being or is likely to be endangered.

11.5   The legislation contemplates three classes of persons to whom disclosure might be made and imposes different rules on motivation in each case. In all cases, the disclosure must be made in good faith to be a qualifying disclosure. The first class of persons to whom disclosure might be made is the worker's employer or, where the worker believes the failure relates solely or mainly to either the conduct of a person other than his employer or a matter for which a person other than his employer has legal responsibility, to such other person. A disclosure to a person under a procedure authorised by the employer is treated as a disclosure to the employer. Disclosures to this class of persons are subject only to the requirement of good faith to qualify.

11.6   A second class of persons to whom disclosure might be made comprises persons prescribed by order of the Secretary of State. The Public Interest Disclosure (Prescribed Persons) Order 1999 names the HSE as a prescribed person in regard to matters which may affect the health and safety of any individual at work, or of any member of the public in connection with the activities of persons at work. The professional construction institutions are not prescribed persons under the Order, nor is the Standing Committee on Structural Safety (SCOSS). To qualify, disclosures to the HSE as a prescribed person are subject to a requirement not only of good faith, but also of reasonable belief that the matters fall within the area for which the HSE is a prescribed person and that the information disclosed, and any allegation contained in it, are substantially true.

11.7 The third and final class of persons to whom disclosures might be made comprises all other persons. Such disclosures are subject to more stringent requirements to qualify. Either the worker must believe that the employer will react adversely if the disclosure was made to him (either by subjecting the worker to a detriment or by concealing or destroying evidence) or the failure must be an exceptionally serious matter. In either case, the disclosure must not only be made in good faith, but the worker must also believe that the information disclosed, or any allegation contained in it, are substantially true, the disclosure must not be made for the purpose of personal gain and, in all the circumstances of the case, it must be reasonable for the worker to make the disclosure. There are factors listed as relevant to whether it is reasonable for the worker to make the disclosure, including the identity of the person to whom the disclosure is made, the seriousness of the relevant failure, and whether the relevant failure is continuing or is likely to occur in the future.

12   General conclusion

12.1   Under current health and safety legislation, the primary responsibility for health and safety lies with 'employers', that is, persons having employees. Under HSWA Section 2 each employer is responsible for the health and safety of his own employees. In addition, under HSWA Section 3, an employer (or any self-employed person) has duties for the health and safety of those who are not his employees, but who are, or could be, affected by his work activities.

12.2   Individuals acting (or purporting to act) as directors or managers or similar officers of an employer organisation may be held personally liable (alongside the organisation itself) for breaches of statutory duties committed by the organisation, if the offence has been committed with their consent or concurrence, or was due to their neglect.

12.3   An employee has a duty under HSWA Section 7 to take care for his own health and safety, and of others who may be affected by his acts or omissions at work, and to co-operate with his employer and any other duty holder.

12.4   Health and safety regulations, made under powers conferred by HSWA Section 15, impose specific duties on employers, employees and the self-employed, and also create specific duty holder roles.

12.5   Most of the duties are based on considering and doing what is reasonably practicable to control risks to health and safety. This entails identifying what risks exist and need to be controlled, and what steps should be taken to manage and control them. Decisions need to be justified by written risk assessments, which are mandatory.

12.6   Historically, under most forms of contract, responsibility for the health and safety of those involved in, and affected by, the work of the construction industry has rested with the contractors, and they have been deemed to have allowed for any consequent costs in their tenders. While contractors and sub-contractors have a continuing role to fulfil, it is clear that there are now responsibilities and legal duties on all those involved in projects. In particular, the CDM Regulations and the CHSWR place duties on all members of project teams according to their functions within the project.

12.7   The legislative developments are essentially based on what is sound practice. The legislation should be interpreted to be consistent with sound practice. Engineering has always been about the identification and control of risks. If all individuals and organisations involved in a project work more closely in the interests of the health and safety of all those they interface with, then improvements will be achieved. This is not just a theory — in parts of the construction industry where clients have taken a strong lead and sensibly provide the necessary resources, health and safety records have shown performance levels significantly better than industry norms.

Appendix A
Rules for Professional Conduct

The Rules for Professional Conduct of all the construction professional institutions include some rules relevant to health and safety. Relevant rules in the *ICE Rules for Professional Conduct* are set out in paragraph 2.1 above. Relevant rules of other institutions are set out below for reference.

IStructE Rules of Conduct (July 1997)

Rule 1. Members of the Institution in their responsibility to the profession shall have full regard to the public interest.

Rule 4. Members of the Institution shall not maliciously or recklessly injure or attempt to injure, whether directly or indirectly, the professional reputation of another engineer.

Rule 6. Members of the Institution shall have a duty to update and broaden their professional knowledge and skills on a continuing basis.

RIBA Code of Professional Conduct (1999)

The Standard of Professional Performance:

Members are required to maintain in their work and that of their practices a standard of performance which is consistent with membership of the RIBA and with a proper regard for the interest both of those who commission and those who may be expected to use or enjoy the product of their work.

Members and their practices will meet the requirements of their engagements with commensurate knowledge and attention so that the quality of the professional services provided does not fall below that which could reasonably be expected of members of the Institute in good standing in the normal conduct of their business.

RICS Conduct and Disciplinary Regulations (2001)

Rule 27.2.8. Every Member shall in the performance of his professional work, the conduct of his practice and the duties of his employment provide the standard of service which it is reasonable to expect of a Chartered Surveyor.

CIOB Rules and Regulations of Professional Competence and Conduct (1993)

Rule 1. Members shall, in fulfilling their professional responsibilities and the duties which they undertake, have full regard to the public interest.

Rule 6. Members shall not undertake work for which they knowingly lack sufficient professional or technical competence, or the adequate resources to meet their obligations.

Rule 12. Members shall not maliciously or recklessly injure or attempt to injure, whether directly or indirectly, the professional reputation, prospects or business of others.

Rule 13. Members shall keep themselves informed of current thinking and developments appropriate to the type and level of their responsibility. They should be able to provide evidence that they have undertaken sufficient study and personal development to fulfil their professional obligations in accordance with the current guidelines for Continuing Professional Development (CPD).

Rule 15. Members shall at all times have due regard for the safety, health and welfare of themselves, colleagues and any others likely to be affected, and in particular be expected to have:

15.1 knowledge of the health and safety risks in the industry and the main principles and strategies for control;

15.2 an understanding of the responsibilities for safety, health and welfare placed on all parties involved in the building process;

15.3 a working knowledge of current legislation and advisory information;

15.4 a recognition of the importance of keeping themselves up-to-date.

Appendix B
The Royal Academy of Engineering Amended Draft Guidelines for Warnings of Preventable Disasters

Offered to the Professional Institutions for Consideration by The Fellowship of Engineering.

SUGGESTED ACTIONS FOR PEOPLE MAKING OR RECEIVING WARNINGS OF DISASTER

1. **Introduction**

1.1   These Guidelines suggest courses of action to assist engineers to react in a responsible, prompt and disciplined manner when they are faced with potentially disastrous situations. Engineers, in the course of their work and at other times, can identify unforeseen risks of disaster to the public or the environment. Others managing public facilities and hazardous installations can be presented with unexpected warnings of potential disaster. Engineers are placed under a professional duty to uphold the safety of the public and the environment by the codes of conduct of their Institutions and organisations. A reciprocal responsibility is placed on the Institutions and organisations to assist any member who turns to them for help in furthering this duty under the code of conduct.

1.2 Well-managed organisations have safety cultures which encourage employees to be vigilant in the identification and elimination of hazardous situations. They encourage employees at all levels to report potentially dangerous situations, and commend the employees even when a warning later proves to have been unfounded. Many organisations have established procedures for making and responding to unexpected warnings; and engineers are expected to work within such procedures where they exist. The systematic reporting of warnings enables newly developing risks to be identified before disaster occurs. These Guidelines may help organisations and Institutions to review their existing lines of communication. However, the Guidelines have been prepared primarily for the non-routine circumstances which occur rarely and which do not fall within established procedures.

1.3 The underlying principle of these Guidelines is that any person needing to make a warning or receiving a warning should draw on his professional peers to verify the risk, decide upon appropriate action and remedy the situation. By sharing the problem the person improves his own credibility and improves the effectiveness of the course of action.

1.4 In the normal course of events a warning can be given, and avoiding actions taken, in an informal manner by the individuals and organisations directly involved. It is anticipated that the more formal procedures of this document will only be followed on very rare occasions.

1.5   This document examines what may be good professional practice in appropriate cases. The Guidelines do not displace or alter the statutory, contractual and civil law duties of the parties involved. The laws in some countries may impose greater duties on organisations and individuals than are implied here.

2.   Actions which might be taken by a person identifying a possible cause of disaster

2.1   Prepare a simple explanation of the potential disaster situation which can be understood by a layman.

2.2   Obtain a second opinion on an advisable course of action from someone competent to understand the failure risk.

2.3   Review your motives for making the warning. Ensure that you could make a declaration that the warning is not influenced by financial or personal considerations.

2.4   Make the warning with explanation to someone in the responsible organisation who is in a position to take action to avoid the possible disaster.

2.5   Enclose a copy of this document with the warning and indicate your availability to discuss the problem.

2.6   Maintain confidentiality.

3.   Actions which might be taken by a person receiving an unsolicited warning of disaster

3.1   Draw the warning to the attention of those ultimately responsible for resolving the situation and obtain a response.

3.2   If the risk of disaster, or the necessity for avoiding action, is not clear cut, obtain a second opinion from a competent person who is truly independent. Guidance may be sought through the Secretary of the appropriate Chartered Engineering Institution.

3.3   Consider your position and, if appropriate, obtain advice on legal liability and implications for insurance cover in the light of the warning received.

3.4   It is desirable that all parties concerned discuss the matter and come to an agreement. If this is not done, advise the person making the warning that action is being taken, or that a second opinion is being obtained.

4.   Notes

4.1   In this document the 'warner' is the person making the warning, while the 'warnee' is the person receiving the warning, and the 'hazard' is the possible disaster situation. 'He' and 'his' should be interpreted as 'she' and 'hers' where appropriate.

4.2   This document is not intended to be exhaustive or restrictive. The course of action in each situation, and the need for detailed calculations and checking, must be decided by the warner and the persons consulted for guidance. Simpler courses of action than those listed may be suitable when the risk or cost of remedy is small, or if effective lines of communication already exist.

4.3   The warner would normally be expected to turn for advice in the first instance to his colleagues or managers. He should continue to obtain guidance from his advisers during subsequent developments. Colleagues and management are likely to understand the repercussions of the problem better than people less familiar with the warner or hazard. The warner may not be able to identify all the factors involved or all the repercussions of a warning. He should take particular care if his concern is not shared by the people he turns to for advice: if he cannot convince friends he is unlikely to persuade others.

4.4   If the warner is an employee, or consultant, of the organisation responsible for resolving the situation, and if he does not resolve the matter quickly with his immediate manager, he should pursue the matter to senior management (preferably with a private meeting). Reference should be made to this document and to the professional duties of his Institution's Code of Conduct.

4.5   It is essential that the warner should retain an attitude of cooperation with the warnee and that he should follow established procedures and lines of communication as far as practicable.

4.6   The warner should make clear whether he is basing his warning on professional knowledge or is acting simply as a non-expert member of the public. If he has professional knowledge relating to the hazard he adds weight to the warning and takes on a greater degree of responsibility, as discussed in this document. If he does not have relevant expertise he should take care not to give the impression that he has.

4.7   Many failures and disasters have resulted from unpredicted oversights or human errors. Moreover, a disaster seldom results from a single cause but rather from a chain of events, the elimination of any one of which may be sufficient to prevent tragedy. Prior to failure the risk can seldom be predicted precisely and, to a certain extent, its assessment is subjective and a matter of opinion. It is therefore important that any second opinion is as objective and independent as possible, and takes account of all the factors considered likely to lead to disaster.

4.8   An engineer may seek advice through the Secretary of the appropriate Engineering Institution on how to proceed and on his professional duties and obligations. He should take care not to disclose the names of other parties or confidential information, unless such disclosure is agreed by the other parties. It is possible that others may wish to turn to the Institution in confidence on the matter.

4.9   Disclosure of confidential information may infringe conditions of employment, which could have serious repercussions for the employment or advancement of the warner. A warner who is concerned about the consequences of a warning on his employment should discuss the matter with his Engineering Institution.

4.10   If an informal warning is not heeded, and the warner and his advisers remain convinced of the seriousness of the hazard, then he should issue a formal written statement to the warnee setting out the reasons why he believes the public or environment is at risk, and indicating how he has followed these Guidelines.

4.11   The warner's obligation to his Institution's Code of Conduct should be discharged by issuing the written statement, except where the warner has in some way contributed to the risk. The warner should seek guidance from the Institution about how much further it is right to take the matter. An employee has no authority to direct his employer, therefore he cannot be held responsible for his employer's conduct. If the employer's action should prove to be detrimental to public health and safety then this would be a matter for adjudication by the Courts.

4.12 If the warner is uncertain as to whom to make the warning within the responsible organisation, he should make the warning to the head of the organisation; e.g. Chairman of the company, Minister of Government Department.

4.13 If the hazard could relate to several organisations and situations that the warner and Engineering Institution may not be able to identify, it may be appropriate to approach a national body, such as the Health and Safety Executive, or Government Department. Under some circumstances the Engineering Institution might consider it appropriate to organise a meeting for discussion which is open to the public. In the case of a generic hazard the Engineering Institution might issue a public warning.

4.14 The warner and the warnee are likely to incur expense which is not recoverable. The warnee, or his advisers, could suffer substantial loss as a consequence of a warning even if they are supported by a second opinion. The warner must take care not to be negligent or careless in communicating the warning. The need for legal advice should be included in the matters discussed with the Engineering Institution.

4.15 The Chartered Engineering Institutions or The Royal Academy of Engineering will endeavor to supply names of appropriate persons and organisations who can provide a second opinion or can undertake an independent safety audit.

4.16   If the disaster does occur the warnee should seek legal advice immediately and the warner should consult his Engineering Institution.

4.17   If this document is used by members of the public The Royal Academy of Engineering will endeavour to advise them on the appropriate organisation to approach for guidance.

4.18   This document is published by The Royal Academy of Engineering solely to assist professional engineers by giving guidance to such engineers about the way they discharge their professional duties in the circumstances described above. The Royal Academy of Engineering hereby expressly disclaims any duty of care, or any other special relationship to any third party and specifically states that it assumes no responsibility or risk at law, however arising, for any use (including the ignoring of any warning) made by any party of these Guidelines and/or any warnings issued because of the existence of these Guidelines.

29 January 1991

With the consent of HM The Queen, The Fellowship of Engineering was renamed The Royal Academy of Engineering in 1992. This extract is reproduced by permission of The Royal Academy of Engineering (www.raeng.org.uk).

References in the original Draft to the Fellowship of Engineering have been amended to refer to The Royal Academy of Engineering except in the title.

Appendix C
Illustration of structure for
goal-setting approach

The recommendation by Robens (1972) that health and safety legislation should adopt a goal-setting approach was revived by Lord Cullen (1990) in his report on the Piper Alpha disaster, in respect of health and safety regulations. Development work carried out in the offshore sector, which was directly affected by the Cullen recommendations, provides a valuable source of ideas for construction generally. The offshore design contractors developed, through their trade organisation (the British Chemical Engineering Contractors' Association (BCECA)), a three-tier structure for the definition of goals. This led to the splitting of technical goals from management goals.

The following chart [Figure 1] illustrates the three-tier structure in the context of managing specific hazards arising from fire and explosion.

- Top tier goal — The Risk Based Performance Standard, a statement of risk targets or limits that the plant or installation intends to achieve. A numerical measure.
- Middle tier goal — The Scenario Based Performance Standard, a description of the hazardous scenario that is introducing the risk. This can be a semantic description

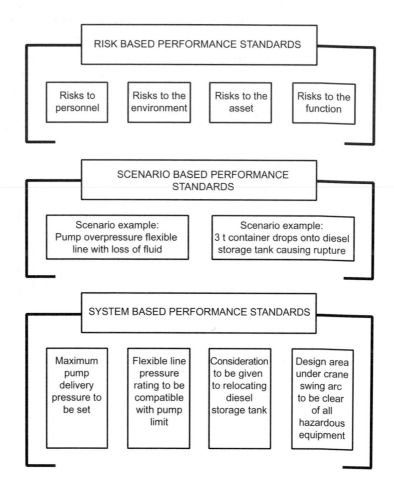

Figure 1. Three-tier structure

of the hazard, but is bespoke to the installation under review.

- Bottom tier goal — The System Based Performance Standard, a statement of the numerical targets that systems/equipment have to achieve to perform their safety function. This should not be confused with specifications. The concept overlaps but is specific to the system/equipment performance in the context of the hazardous scenario described in the higher tier goal above.

[The figure and text in this Appendix are reproduced from a paper by Edmund Terry and Simon Dean, The importance of design in achieving health and safety: lessons from the offshore industry, in *Designing for Safety and Health, Conference Proceedings* (A. Gibb (ed.)), European Construction Institute, 2000, ISBN 1873 844 48 4.]

References

EU Directives

Health and Safety Framework Directive (89/391/EEC)

Temporary or Mobile Construction Sites Directive (92/57/EEC)

(EU Directives are available at www.europa.eu.int/eur-lex)

Statutes

Contracts (Rights of Third Parties) Act 1999

Employers' Liability (Compulsory Insurance) Act 1969

Employment Rights Act 1996

Health and Safety at Work etc. Act 1974

Limited Liability Partnerships Act 2000

Occupier's Liability Act 1957

Occupier's Liability Act 1984

Occupier's Liability (Scotland) Act 1960

Public Interest Disclosure Act 1998

Unfair Contract Terms Act 1977

(The texts of Acts of Parliament since 1988 are available, as originally enacted, at www.legislation.hmso.gov.uk/acts.htm. This does not, however, show amendments.)

Statutory Instruments

Construction (Design and Management) Regulations 1994 (SI 1994/3140)

Construction (Design and Management) (Amendment) Regulations 2000 (SI 2000/2380)

Confined Spaces Regulations 1997 (SI 1997/1713)

Construction (Head Protection) Regulations 1989 (SI 1989/2209) (as amended by Part X of the Personal Protection Equipment Regulations 1992 and 1999)

Construction (Health, Safety and Welfare) Regulations 1996 (SI 1996/1592)

Control of Asbestos at Work Regulations 1987 (SI 1987/2115)

Control of Lead at Work Regulations 1998 (SI 1998/543)

Control of Major Accident Hazards Regulations 1999 (SI 1999/743)

Control of Substances Hazardous to Health Regulations 1999 (SI 1999/437)

Diving at Work Regulations 1997 (SI 1997/2776)

Docks Regulations 1988 (SI 1988/1655)

Fire Precautions (Workplace) Regulations 1997 (SI 1997/1840) (as amended by SI 1999/1877 and SI 1999/3242)

Health and Safety (First Aid) Regulations 1981 (SI 1981/917)

Lifting Operations and Lifting Equipment Regulations 1998 (SI 1998/2307)

Management of Health and Safety at Work Regulations 1999 (SI 1999/3242)

Manual Handling Operations Regulations 1992 (SI 1992/2793)

Noise at Work Regulations 1989 (SI 1989/1790)

Personal Protective Equipment Regulations 1992/99 (SI 1992/2966) (as amended by SI 1999/860 and SI 1999/3232)

Pipelines Safety Regulations 1996 (SI 1996/825)

Provision and Use of Work Equipment Regulations 1998 (SI 1998/2306)

Public Interest Disclosure (Prescribed Persons) Order 1999 (SI 1999/1549)

Quarries Miscellaneous Health and Safety Provisions Regulations 1995 (SI 1995/2036)

Railways (Safety Case) Regulations 1994 (SI 1994/237)

Reporting of Injuries, Diseases and Dangerous Occurrences Regulations 1995 (SI 1995/3163)

Supply of Machinery (Safety) Regulations 1992 (SI 1992/3073)

Work in Compressed Air Regulations 1996 (SI 1996/1656)

Workplace (Health, Safety and Welfare) Regulations 1992 (SI 1992/3004)

(The texts of Statutory Instruments since 1987, as originally published, are available at www.legislation.hmso.gov.uk/stat.htm. The text of Regulations is also included in the relevant ACoPs.)

Cases

AMF International Ltd v *Magnet Bowling Ltd* [1968] 1 WLR 1028; [1968] 2 All ER 789

R v *Associated Octel Co. Ltd* [1996] 4 All ER 846

R v *Associated Octel Co. Ltd* [1997] 1 Cr App R(S) 435 (re costs only)

R v *Cardiff City Transport Services* [2001] 1 Cr App R(S) 141

R v *Howe* [1999] 2 All ER 249; [1999] 2 Cr App R(S) 37

R v *Rollco Screw and Rivet Co. and others* [1999] 2 Cr App R(S) 436

R v *Swan Hunter* [1982] 1 All ER 264

Ramsgate Walkway Trial (Unreported). For relevant extracts from the judgment see: J. Barber, Ramsgate Walkway Collapse: Legal

Ramifications, in *Forensic Engineering*, Thomas Telford Publishing, 1999 (also in [2001] 17 Const LJ 25).

HSE and other Government publications

Lord Cullen, *Public Inquiry into the Piper Alpha Disaster* 1990 Cmnd 1310

Health and Safety Executive, *Health and Safety in Construction* HSG 150 (rev), HSE Books, 2001, ISBN 0 7176 2106 5

Health and Safety Executive, *The Collapse of NATM Tunnels at Heathrow Airport*, HSE Books, 2000, ISBN 0 7176 1792 0

Health and Safety Executive, *Management of health and safety at work: Management of Health and Safety at Work Regulations 1999 Approved Code of Practice and Guidance*, 2000, HSE Books, ISBN 0 7176 2488 9

Health and Safety Executive, *Managing Health and Safety in Construction: Construction (Design and Management) Regulations 1994 Approved Code of Practice and Guidance*, HSE Books, 2001, ISBN 0 7176 2139 1

Health and Safety Executive, *Reducing risks protecting people — HSE's decision-making process*, HSE Books, 2001, ISBN 0 7176 2151 0

Lord Robens, *Safety and Health at Work, Report of the Committee 1970-72* Cmnd 5034

(Obtainable from HSE Books at www.hsebooks.co.uk, tel. no. 01787 881165; or from The Stationery Office.)

Other Publications

J. Anderson, Construction Safety — time for a legislative tidy up, *Proceedings of the Institution of Civil Engineers, Civil Engineering*, 2001, **144**, Feb., p. 4

Construction Industry Research and Information Association, *Safety in ports — ship-to-shore linkspans and walkways, A guide to procurement, operation and maintenance*, CIRIA, 1999, CIRIA Report C518, ISBN 0 8617 518 9

In addition to the printed versions and the websites detailed above, relevant EU Directives, Statutes, Statutory Instruments (all as amended to date) and Cases are available on CD-Rom from Butterworths' Health and Safety Law Service (with Redgrave's Health and Safety — visit www.butterworths.co.uk for details) or from Gee Publishing (also with ACoPs) as a subscription web-based source at www.safety-now.co.uk.

No references have been included for superseded legislation or publications.